A P P E N D I X C

# Materials and Process Selection Charts

## C1  Introduction

This appendix is a compilation of the Materials and Process Selection Charts, for use in problem solving. Each chart appears on a single page with a brief commentary about its use on the facing page. Background details are given in Chapters 4 and 14 for the Materials Charts, and in Chapter 9 for the Process Charts.

The Materials Charts are most effectively used by plotting performance indices onto them, isolating a subset of materials which optimally meet the design goals. A number of these indices are derived in the text and summarised in Tables 5.1 and 7.3. Examples of their use are given in Chapters 6, 8 and 14. Process Charts are used by plotting process requirements onto them, isolating a subset of viable processes, as described in Chapters 9 and 10. Sequential application of several charts allows several design goals to be met simultaneously. More advanced methods are described in Section 5.4.

The best way to tackle selection problems is to work directly on the appropriate charts. Although the book itself is copyright, the reader is authorised to make copies of the charts, and to reproduce these, with proper reference to their source, as he or she wishes.

### Cautions

The data on the charts and in the tables are approximate: they typify each class of material (stainless steels, or polyethylenes, for instance), but within each class there is considerable variation. They are adequate for the broad comparisons required in *conceptual* design, and, often, for the rough calculations of *embodiment* design. *THEY ARE NOT APPROPRIATE FOR DETAILED DESIGN CALCULATIONS*. For these, it is essential to seek accurate data from handbooks and the data sheets provided by material suppliers. The charts help in narrowing the choice of candidate materials to a sensible short list, but not in providing numbers for final accurate analysis.

Every effort has been made to ensure the accuracy of the data shown on the charts. The sources are detailed in Chapters 4, 9 and 11. No guarantee can, however, be given that the data are error-free, or that new data may not supersede that given here. The charts are an aid to creative thinking, not a source of numerical data for precise analysis.

### C2  Material Classes and Class Members

The materials of mechanical and structural engineering fall into nine broad classes.

TABLE 1. *Material Classes*

| | |
|---|---|
| Engineering Alloys | (Metals and their alloys) |
| Engineering Polymers | (Thermoplastics and thermosets) |
| Engineering Ceramics | ("Fine" ceramics) |
| Engineering Composites | (GFRP, KFRP and CFRP) |
| Porous Ceramics | (Brick, cement, concrete, stone) |
| Glasses | (Silicate glasses) |
| Woods | (Common structural timbers) |
| Elastomers | (Natural and artificial rubbers) |
| Foams | (Foamed polymers) |

Within each class, the Materials Selection Charts show data for a representative set of materials, chosen both to span the full range of behaviour for that class, and to include the most widely used members of it. In this way the envelope for a class (heavy lines) encloses data not only for the materials listed on Table 2, but for virtually all other members of the class as well.

As far as possible, the same materials appear on all the charts. There are exceptions. Invar is only interesting because of its low thermal expansion: it appears on the thermal expansion charts (10 and 11) but on no others. Mu − Cu alloys have high internal damping: they are shown on the loss coefficient chart (8), but not elsewhere. And there are others. But, broadly, the materials and classes which appear on one chart appear on them all.

You will *not* find specific materials listed on the charts. The aluminium alloy 7075 in the T6 condition (for instance) is contained in the property envelopes for *Al-alloys*; the Nylon 66 in those for *nylons*. The charts are designed for the broad, early stages of materials selection, not for retrieving the precise values of material properties needed in the later, detailed design, stage. Procedures for finding information at this higher level of accuracy are described in Chapters 11 and 12.

### C3  Process Classes and Class Members

The manufacturing processes of engineering fall into nine broad classes:

TABLE 3. *Process Classes*

| | |
|---|---|
| Casting | (Sand, Gravity, Pressure, Die, etc.) |
| Pressure moulding | (Direct, Transfer, Injection, etc.) |
| Deformation processes | (Rolling, Forging, Drawing, etc.) |
| Powder methods | (Slip cast, Sinter, Hot press, HIP) |
| Special methods | (CVD, Electroform, Lay-up, etc.) |
| Machining | (Cut, Turn, Drill, Mill, Grind, etc.) |
| Heat treatment | (Quench, Temper, Solution treat, Age, etc.) |
| Joining | (Bolt, Rivet, Weld, Braze, Adhesives) |
| Finishing | (Polish, Plate, Anodise, Paint) |

TABLE 2. *Material Classes and Members of Each Class*

| Class | Members | Short Name |
|---|---|---|
| Engineering Alloys (The metals and alloys of engineering) | Aluminium alloys | Al alloys |
| | Copper alloys | Cu alloys |
| | Lead alloys | Lead alloys |
| | Magnesium alloys | Mg alloys |
| | Molybdenum alloys | Mo alloys |
| | Nickel alloys | Ni alloys |
| | Steels | Steels |
| | Tin alloys | Tin alloys |
| | Titanium alloys | Ti alloys |
| | Tungsten alloys | W alloys |
| | Zinc alloys | Zn alloys |
| Engineering Polymers (The thermoplastics and thermosets of engineering) | Epoxies | EP |
| | Melamines | MEL |
| | Polycarbonate | PC |
| | Polyesters | PEST |
| | Polyethylene, high density | HDPE |
| | Polyethylene, low density | LDPE |
| | Polyformaldehyde | PF |
| | Polymethylmethacrylate | PMMA |
| | Polypropylene | PP |
| | Polytetrafluorethylene | PTFE |
| | Polyvinylchloride | PVC |
| Engineering Ceramics (Fine ceramics capable of load-bearing application) | Alumina | $Al_2O_3$ |
| | Diamond | C |
| | Sialons | Sialons |
| | Silicon carbide | SiC |
| | Silicon nitride | $Si_3N_4$ |
| | Zirconia | $ZrO_2$ |
| Engineering Composites (The composites of engineering practice). A distinction is drawn between the properties of a ply — "UNIPLY" — and of a laminate — "LAMINATES" | Carbon fibre reinforced polymer | CFRP |
| | Glass fibre reinforced polymer | GFRP |
| | Kevlar fibre reinforced polymer | KFRP |
| Porous Ceramics (Traditional ceramics, cements, rocks and minerals) | Brick | Brick |
| | Cement | Cement |
| | Common rocks | Rocks |
| | Concrete | Concrete |
| | Porcelain | Pcln |
| | Pottery | Pot |
| Glasses (Ordinary silicate glass) | Borosilicate glass | B-glass |
| | Soda glass | Na-glass |
| | Silica | $SiO_2$ |
| Woods (Separate envelopes describe properties parallel to the grain and normal to it, and wood products) | Ash | Ash |
| | Balsa | Balsa |
| | Fir | Fir |
| | Oak | Oak |
| | Pine | Pine |
| | Wood products (ply, etc.) | Wood products |
| Elastomers (Natural and artificial rubbers) | Natural rubber | Rubber |
| | Hard butyl rubber | Hard butyl |
| | Polyurethanes | PU |
| | Silicone rubber | Silicone |
| | Soft butyl rubber | Soft butyl |
| Polymer Foams (Foamed polymers of engineering) | These include: | |
| | Cork | Cork |
| | Polyester | PEST |
| | Polystyrene | PS |
| | Polyurethane | PU |

The Process Selection Charts show the ranges of size, shape, material, precision and surface finish of which each class of process is capable.

They are used in the way described in Chapters 9 and 10. The procedure does not lead to a final choice of process. Instead, it identifies a subset of processes which have the potential to meet the design requirements. More specialised sources must then be consulted to determine which of these is the most economical.

# C4   The Materials Selection Charts

### Materials Selection Chart 1

### Young's Modulus, E, against Density, ρ

The chart guides selection of materials for light, stiff, components. The contours show the longitudinal wave speed in m/s; natural vibration frequencies are proportional to this quantity. The guidelines show the loci of points for which

(a)  $E/\rho = C$ (minimum weight design of stiff ties; minimum deflection in centrifugal loading, etc.);

(b)  $E^{1/2}/\rho = C$ (minimum weight design of stiff beams, shafts and columns);

(c)  $E^{1/3}/\rho = C$ (minimum weight design of stiff plates).

The value of the constant $C$ increases as the lines are displaced upwards and to the left. Materials offering the greatest stiffness-to-weight ratio lie towards the upper left corner. Other moduli are obtained approximately from $E$ using

(a)  $\nu \approx \dfrac{1}{3}$ ; $G \approx \dfrac{3}{8} E$; $K \approx E$ (metals, ceramics, glass and glassy polymers);

(b)  $\nu \approx \dfrac{1}{2}$ ; $G \approx \dfrac{1}{3}E$ ; $K \approx 10\,E$ (elastomers, rubbery polymers);

where $\nu$ is Poisson's ratio, $G$ the shear modulus and $K$ the bulk modulus.

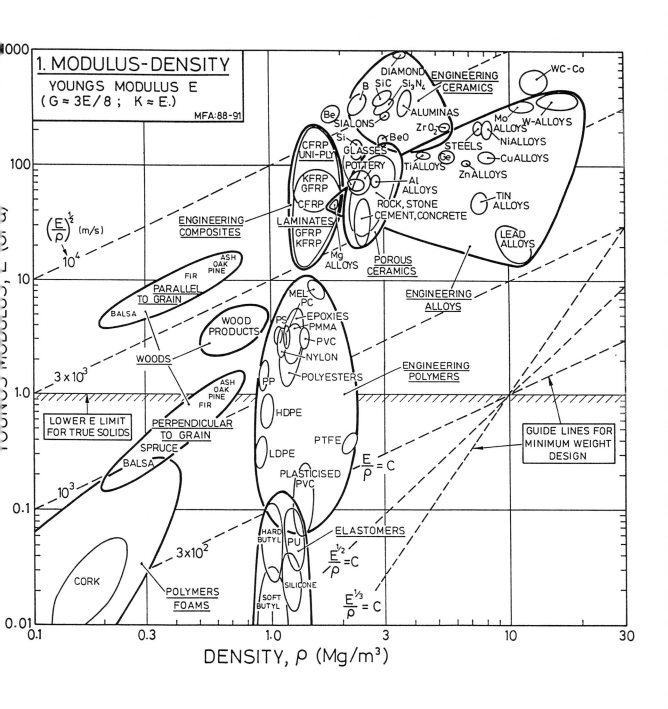

# 1. MODULUS-DENSITY

## YOUNGS MODULUS E
$(G \approx 3E/8; \quad K \approx E.)$

MFA:88-91

### Materials Selection Chart 2

### Strength, $\sigma_f$, against Density, $\rho$

The "strength" for *metals* is the 0.2% offset yield strength. For *polymers*, it is the stress at which the stress-strain curve becomes markedly nonlinear — typically, a strain of about 1%. For *ceramics and glasses*, it is the compressive crushing strength; remember that this is roughly fifteen times larger than the tensile (fracture) strength. For *composites*, it is the tensile strength. For *elastomers*, it is the tear strength. The chart guides selection of materials for light, strong, components. The guidelines show the loci of points for which

(a)   $\sigma_f/\rho = C$ (minimum weight design of strong ties; maximum rotational velocity of disks);

(b)   $\sigma_f^{2/3}/\rho = C$ (minimum weight design of strong beams and shafts);

(v)   $\sigma_f^{1/2}/\rho = C$ (minimum weight design of strong plates).

The value of the constant $C$ increases as the lines are displaced upwards and to the left. Materials offering the greatest strength-to-weight ratio lie towards the upper left corner.

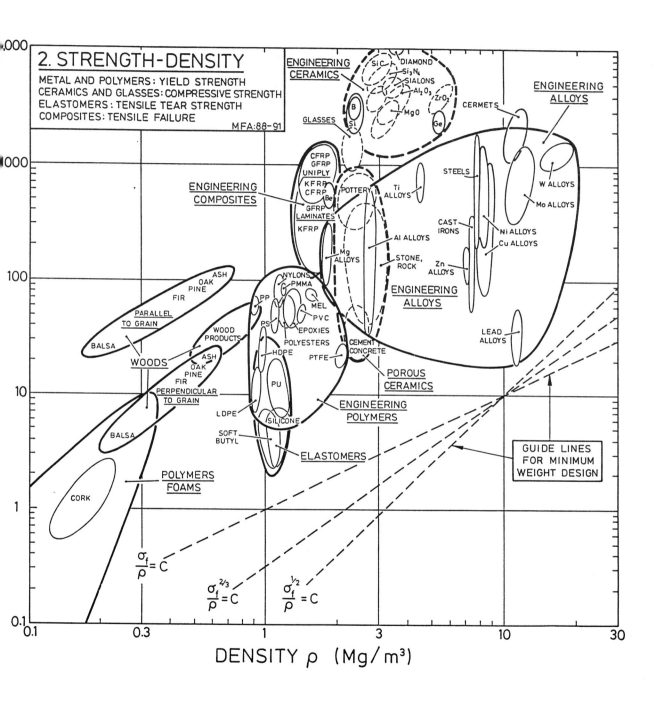

# 2. STRENGTH-DENSITY

METAL AND POLYMERS: YIELD STRENGTH
CERAMICS AND GLASSES: COMPRESSIVE STRENGTH
ELASTOMERS: TENSILE TEAR STRENGTH
COMPOSITES: TENSILE FAILURE

MFA:88-91

DENSITY $\rho$ (Mg/m³)

## Materials Selection Chart 3

### Fracture Toughness, $K_{Ic}$, against Density, $\rho$

Linear elastic fracture mechanics describes the behaviour of cracked, brittle solids. It breaks down when the fracture toughness is large and the section is small; then *J*-integral methods should be used. The data shown here are adequate for the rough calculations of conceptual design and as a way of ranking materials. The chart guides selection of materials for light, fracture-resistant components. The guidelines show the loci of points for which

(a) $\quad K_{Ic}^{4/3}/\rho = C$ $\qquad$ (minimum weight design of brittle ties, maximum
$\quad\ K_{Ic}/\rho = C$ $\qquad\qquad$ rotational velocity of brittle disks, etc.);

(b) $\quad K_{Ic}^{4/5}/\rho = C$ $\qquad$ (minimum weight design of brittle beams
$\quad\ K_{Ic}^{2/3}/\rho = C$ $\qquad\qquad$ and shafts);

(c) $\quad K_{Ic}^{2/3}/\rho = C$ $\qquad$ (minimum weight design of brittle
$\quad\ K_{Ic}^{1/2}/\rho = C$ $\qquad\qquad$ plates).

The value of the constant $C$ increases as the lines are displaced upwards and to the left. Materials offering the greatest toughness-to-weight ratio lie towards the upper left corner.

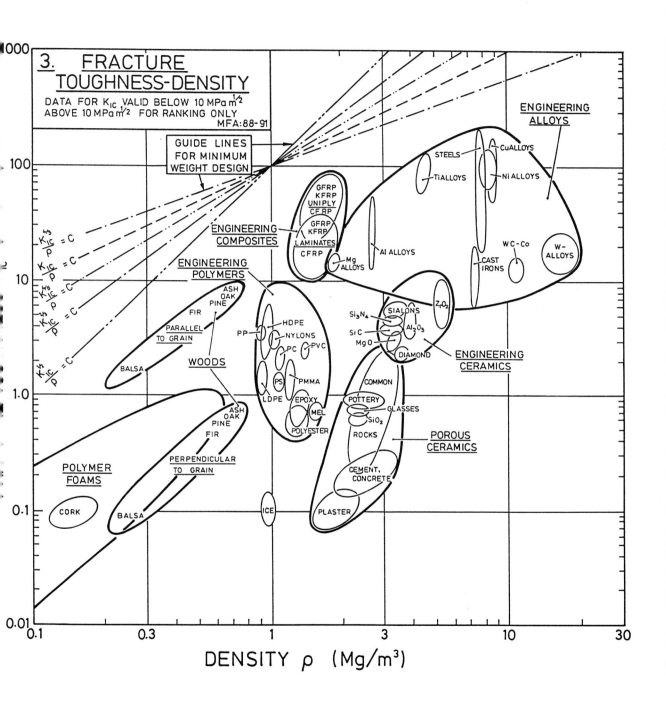

### 3. FRACTURE TOUGHNESS-DENSITY

DATA FOR $K_{IC}$ VALID BELOW 10 MPa m$^{1/2}$
ABOVE 10 MPa m$^{1/2}$ FOR RANKING ONLY

MFA:88-91

GUIDE LINES FOR MINIMUM WEIGHT DESIGN

$\frac{K_{IC}}{\rho} = C$

$\frac{K_{IC}^{2/3}}{\rho} = C$

$\frac{K_{IC}^{1/2}}{\rho} = C$

ENGINEERING ALLOYS

ENGINEERING COMPOSITES

ENGINEERING POLYMERS

ENGINEERING CERAMICS

POROUS CERAMICS

WOODS

PARALLEL TO GRAIN

PERPENDICULAR TO GRAIN

POLYMER FOAMS

STEELS · CuALLOYS
TiALLOYS · NiALLOYS
WC-Co · W-ALLOYS
CAST IRONS
Al ALLOYS
Mg ALLOYS
$Zr O_2$
$Si_3N_4$ · SIALONS
SiC · $Al_2O_3$
MgO · DIAMOND

GFRP KFRP UNIPLY CFRP
GFRP KFRP LAMINATES CFRP

ASH OAK PINE
FIR
BALSA
ASH OAK PINE
FIR
BALSA

HDPE
PP · NYLONS
PC · PVC
PS · PMMA
LDPE · EPOXY
MEL
POLYESTER

COMMON
POTTERY · GLASSES
$SiO_2$
ROCKS

CEMENT, CONCRETE

PLASTER

CORK

ICE

**DENSITY $\rho$ (Mg/m³)**

0.1    0.3    1    3    10    30

1000    100    10    1.0    0.1    0.01

## Materials Selection Chart 4
## Young's Modulus, E, against Strength, $\sigma_f$

The chart for elastic design. The contours show the failure strain, $\sigma_f/E$. The "strength" for *metals* is the 0.2% offset yield strength. For *polymers*, it is the 1% yield strength. For *ceramics* and *glasses*, it is the compressive crushing strength; remember that this is roughly fifteen times larger than the tensile (fracture) strength. For *composites*, it is the tensile strength. For *elastomers*, it is the tear strength. The chart has numerous applications, among them: the selection of materials for springs, elastic hinges, pivots and elastic bearings, and for yield-before-buckling design. The guidelines show three of these; they are the loci of points for which

(a)     $\sigma_f^2/E$   $= C$ (elastic energy storage per unit volume);
(b)     $\sigma_f^{3/2}/E$   $= C$ (selection for elastic constants such as knife edges; elastic diaphragms);
(c)     $\sigma_f/E$   $= C$ (seals, elastic hinges).

The value of the constant $C$ increases as the lines are displaced downward and to the right.

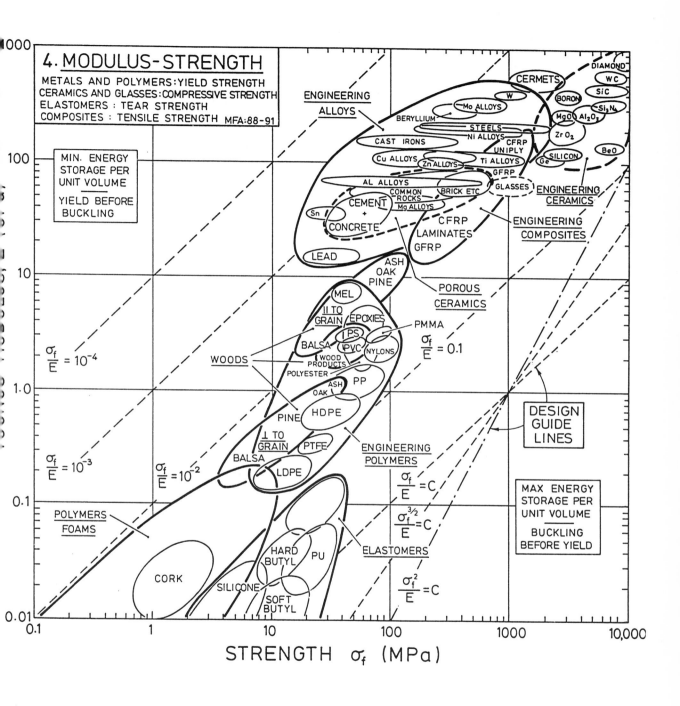

## Materials Selection Chart 5

### Specific Modulus, $E/\rho$, against Specific Strength, $\sigma_f/\rho$

The chart for specific stiffness and strength. The contours show the yield strain, $\sigma_f/E$. The qualifications on strength given for Charts 2 and 4 apply here also. The chart finds application in minimum weight design of ties and springs, and in the design of rotating components to maximise rotational speed or energy storage, etc. The guidelines show the loci of points for which

(a)   $\sigma_f^2/E\rho = C$ (ties, springs of minimum weight; maximum rotational velocity of disks);
(b)   $\sigma_f^{3/2}/E\rho^{1/2} = C$;
(c)   $\dfrac{\sigma_f}{E} = C$ (elastic hinge design).

The value of the constant $C$ increases as the lines are displaced downwards and to the right.

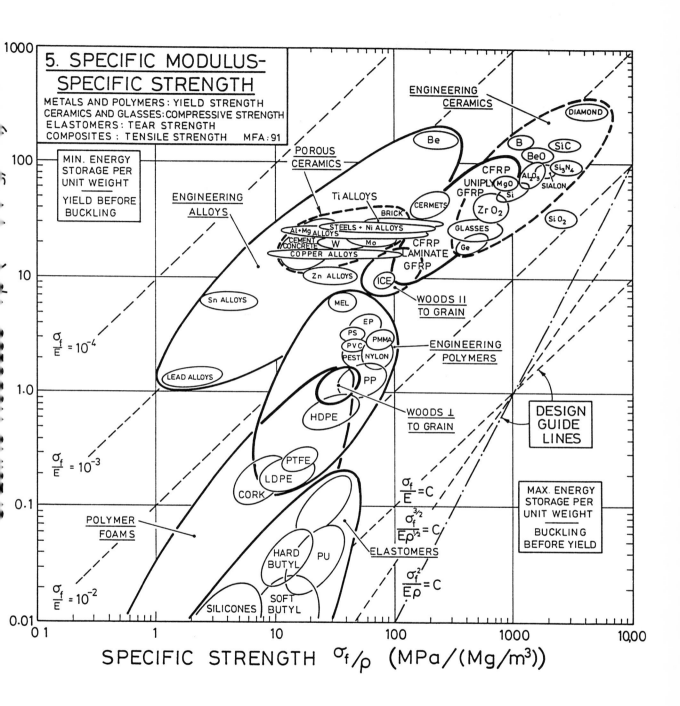

## 5. SPECIFIC MODULUS-SPECIFIC STRENGTH

METALS AND POLYMERS: YIELD STRENGTH
CERAMICS AND GLASSES: COMPRESSIVE STRENGTH
ELASTOMERS: TEAR STRENGTH
COMPOSITES: TENSILE STRENGTH    MFA:91

SPECIFIC STRENGTH $\sigma_f/\rho$ (MPa/(Mg/m³))

## Materials Selection Chart 6

### Fracture toughness, $K_{Ic}$, against Young's Modulus, E

The chart displays both the fracture toughness, $K_{Ic}$, and (as contours) the toughness, $G_{Ic} \approx K^2{}_{Ic}/E$; and it allows criteria for stress and displacement-limited failure criteria ($K_{Ic}$ and $K_{Ic}/E$) to be compared. The guidelines show the loci of points for which

(a)  $K^2{}_{Ic}/E = C$ (lines of constant toughness, $G_c$; energy-limited failure):
(b)  $K_{Ic}/E = C$ (guideline for displacement-limited brittle failure).

The value of the constant $C$ increases as the lines are displaced upwards and to the left. Tough materials lie towards the upper left corner, brittle materials towards the bottom right.

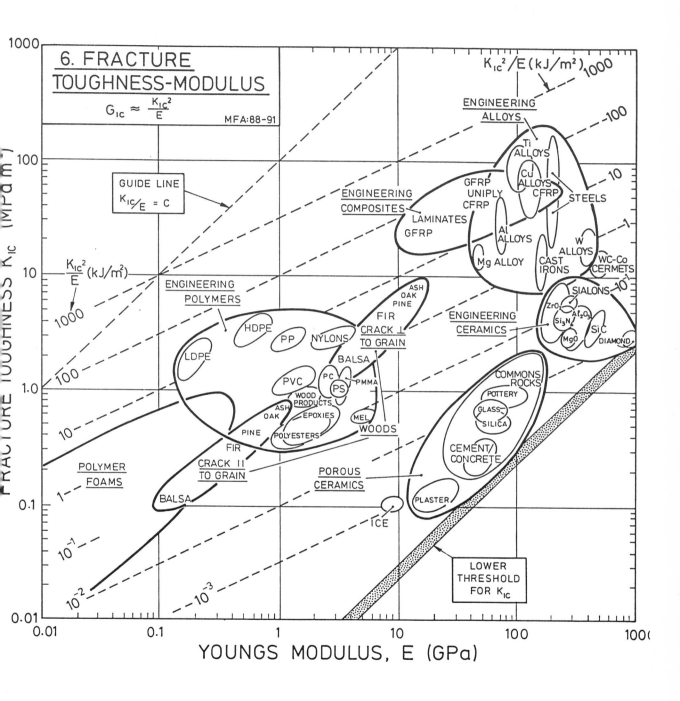

# 6. FRACTURE TOUGHNESS-MODULUS

$$G_{IC} \approx \frac{K_{IC}^2}{E}$$

MFA:88-91

### *Materials Selection Chart 7*

### *Fracture Toughness, $K_{Ic}$, against Strength, $\sigma_f$*

The chart for safe design against fracture. The contours show the process zone diameter, given approximately by $K_{Ic}^2/\pi\sigma_f^2$. The qualifications on "strength" given for Charts 2 and 4 apply here also. The chart guides selection of materials to meet yield-before-break and leak-before-break design criteria, in assessing plastic or process zone sizes, and in designing samples for valid fracture toughness testing. The guidelines show the loci of points for which

(a)   $K_{Ic}/\sigma_f = C$ (yield-before-break);

(b)   $K_{Ic}^2/\sigma_f = C$ (leak-before-break).

The value of the constant $C$ increases as the lines are displaced upward and to the left.

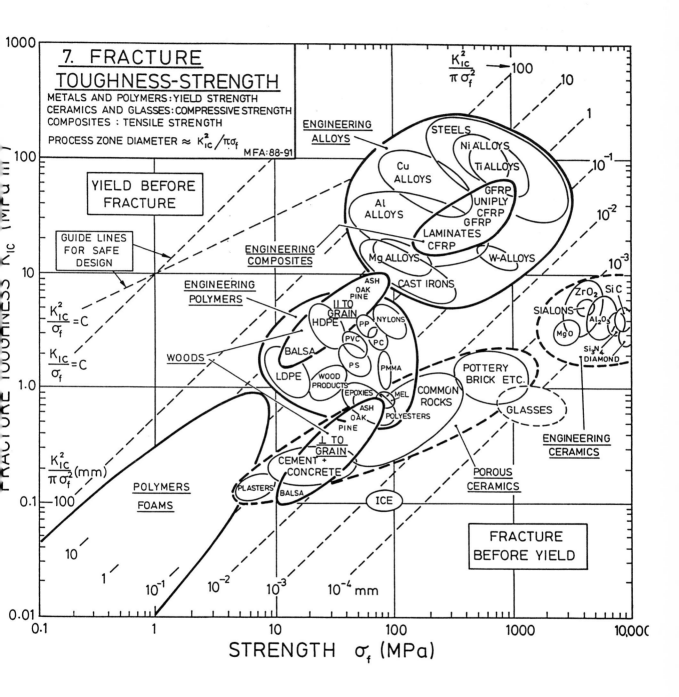

### 7. FRACTURE TOUGHNESS-STRENGTH

METALS AND POLYMERS: YIELD STRENGTH
CERAMICS AND GLASSES: COMPRESSIVE STRENGTH
COMPOSITES : TENSILE STRENGTH

PROCESS ZONE DIAMETER $\approx K_{IC}^2/\pi\sigma_f$

MFA:88-91

$\dfrac{K_{IC}^2}{\pi\sigma_f^2} \longrightarrow$ 100

YIELD BEFORE FRACTURE

GUIDE LINES FOR SAFE DESIGN

$\dfrac{K_{IC}^2}{\sigma_f} = C$

$\dfrac{K_{IC}}{\sigma_f} = C$

ENGINEERING ALLOYS

ENGINEERING COMPOSITES

ENGINEERING POLYMERS

WOODS

ENGINEERING CERAMICS

STEELS
Ni ALLOYS
Ti ALLOYS
Cu ALLOYS
Al ALLOYS
GFRP
UNIPLY CFRP
GFRP
LAMINATES CFRP
Mg ALLOYS
W-ALLOYS
CAST IRONS

ASH
OAK
PINE
II TO GRAIN
HDPE
NYLONS
PP
PVC
PC
BALSA
PS
PMMA
LDPE
WOOD PRODUCTS
EPOXIES
MEL
POLYESTERS
ASH
OAK
PINE
⊥ TO GRAIN
CEMENT + CONCRETE

COMMON ROCKS

POTTERY BRICK ETC.

GLASSES

ZrO₂ → $ZrO_2$  SiC → Si C
SIALONS
Al₂O₃ → $Al_2O_3$
MgO → Mg O
Si₃N₄ → $Si_3N_4$
DIAMOND

$\dfrac{K_{IC}^2}{\pi\sigma_f^2}$ (mm)

POLYMERS FOAMS

PLASTERS
BALSA
ICE

PROUS CERAMICS

FRACTURE BEFORE YIELD

10   1   $10^{-1}$   $10^{-2}$   $10^{-3}$   $10^{-4}$ mm

**STRENGTH** $\sigma_f$ (MPa)

## Materials Selection Chart 8

### Loss Coefficient, $\eta$, against Young's Modulus, E

The chart gives guidance in selecting material for low damping (springs, vibrating reeds, etc.) and for high damping (vibration mitigating systems). The guideline shows the loci of points for which

(a)   $\eta E = C$ (rule-of-thumb for estimating damping in polymers).

The value of the constant $C$ increases as the line is displaced upward and to the right.

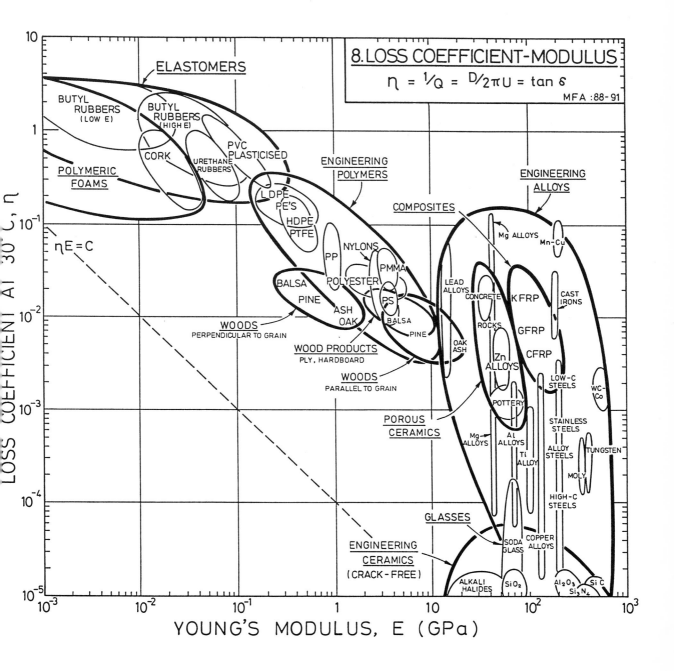

**8. LOSS COEFFICIENT-MODULUS**

$$\eta = {}^1/_Q = {}^D/_{2\pi U} = \tan \delta$$

MFA :88-91

## Materials Selection Chart 9

## Thermal Conductivity, $\lambda$, against Thermal Diffusivity, a

The chart guides in selecting materials or thermal insulation, for use as heat sinks and so forth, both when heat flow is steady ($\lambda$) and when it is transient ($a = \lambda/\rho C_p$ where $\rho$ is the density and $C_p$ the specific heat). Contours show values of the volumetric specific heat, $\rho C_p = \lambda/a$ (J/m$^3$ K). The guidelines show the loci of points for which

(a)  $\lambda/a$  $= C$ (constant volumetric specific heat);

(b)  $\lambda/a^{1/2} = C$ (efficient insulation; thermal energy storage).

The value of the constant $C$ increases towards the upper left.

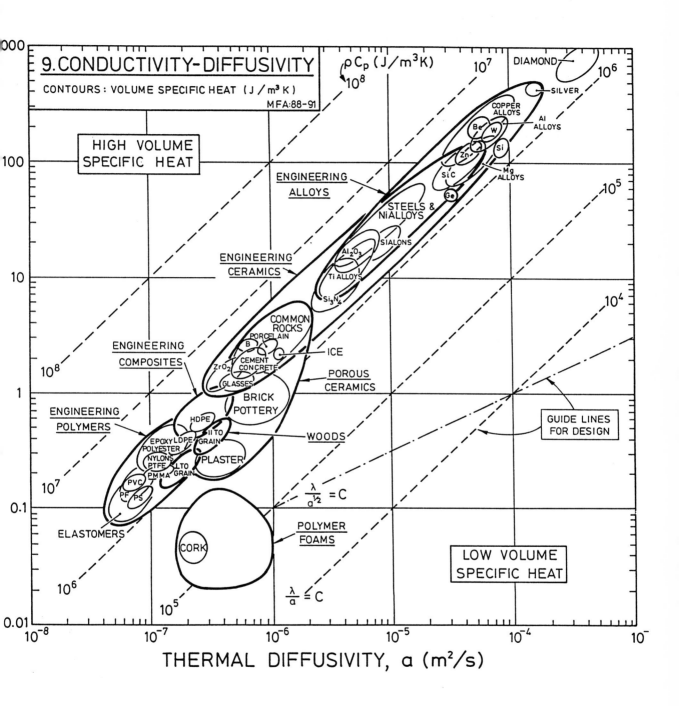

**9. CONDUCTIVITY-DIFFUSIVITY**

CONTOURS: VOLUME SPECIFIC HEAT (J / m³ K)

MFA:88-91

$\rho\, C_p$ ( J / m³K )

HIGH VOLUME SPECIFIC HEAT

ENGINEERING ALLOYS

ENGINEERING CERAMICS

ENGINEERING COMPOSITES

ENGINEERING POLYMERS

ELASTOMERS

DIAMOND

SILVER

COPPER ALLOYS

Al ALLOYS

Be   W

Zn   Si

SiC   Mg ALLOYS

Ge

STEELS & NiALLOYS

Al₂O₃

SIALONS

Ti ALLOYS

Si₃N₄

COMMON ROCKS

PORCELAIN

B

ICE

ZrO₂   CEMENT CONCRETE

GLASSES

BRICK POTTERY

POROUS CERAMICS

HDPE

EPOXY  LDPE

POLYESTER   II TO GRAIN

NYLONS

PTFE   I TO

PMMA   GRAIN

PVC

PF  PS

PLASTER

WOODS

GUIDE LINES FOR DESIGN

POLYMER FOAMS

CORK

LOW VOLUME SPECIFIC HEAT

$\dfrac{\lambda}{a^{\frac{1}{2}}} = C$

$\dfrac{\lambda}{a} = C$

10⁸   10⁷   10⁶   10⁵   10⁴

10⁸   10⁷   10⁶   10⁵

**THERMAL DIFFUSIVITY, a (m²/s)**

### Materials Selection Chart 10
### Thermal Expansion Coefficient, $\alpha$, against Thermal Conductivity, $\lambda$

The chart for assessing thermal distortion. The contours show value of the ratio $\lambda/\alpha$ (W/m). Materials with a large value of this design index show small thermal distortion. They define the guideline

(a)      $\lambda/\alpha = C$ (minimisation of thermal distortion).

The value of the constant $C$ increases towards the bottom right.

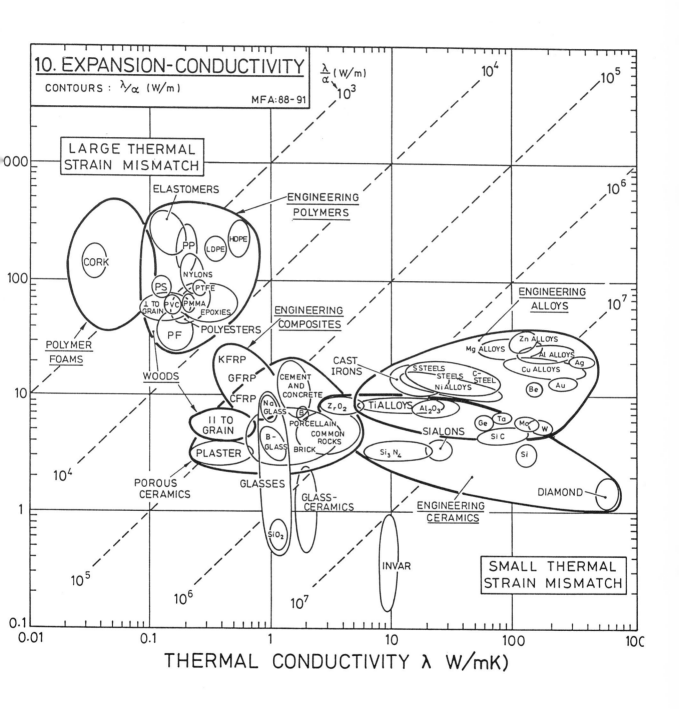

## 10. EXPANSION-CONDUCTIVITY

CONTOURS : $\lambda/\alpha$ (W/m)

MFA:88-91

$\frac{\lambda}{\alpha}$ (W/m)

LARGE THERMAL STRAIN MISMATCH

ELASTOMERS

ENGINEERING POLYMERS

PP    LDPE    HDPE

CORK

NYLONS

PS    PTFE

$\perp$ TO GRAIN    PVC    PMMA    EPOXIES

POLYMER FOAMS

PF    POLYESTERS

ENGINEERING COMPOSITES

WOODS

KFRP

GFRP

CFRP

CEMENT AND CONCRETE

CAST IRONS

ENGINEERING ALLOYS

Mg ALLOYS    Zn ALLOYS    Al ALLOYS    Ag

S STEELS    STEELS    C-STEEL

Ni ALLOYS    Cu ALLOYS

Be    Au

$\parallel$ TO GRAIN

Na GLASS

B A

PORCELLAIN

$Z_rO_2$    TiALLOYS    $Al_2O_3$

PLASTER

B-GLASS    COMMON ROCKS    BRICK

SIALONS

Ge    Ta    Mo    W

SiC

Si

POROUS CERAMICS

GLASSES

GLASS-CERAMICS

$Si_3N_4$

ENGINEERING CERAMICS

DIAMOND

$SiO_2$

INVAR

SMALL THERMAL STRAIN MISMATCH

$10^3$    $10^4$    $10^5$

$10^6$

$10^7$

$10^4$

$10^5$    $10^6$    $10^7$

**THERMAL CONDUCTIVITY $\lambda$ W/mK)**

## Materials Selection Chart 11

### Linear Thermal Expansion, $\alpha$, against Young's Modulus, E

The chart guides in selecting materials when thermal stress is important. The contours show the thermal stress generated, per °C temperature change, in a constrained sample. They define the guideline

(a)   $\alpha E = C$ MPa/K (constant thermal stress per °K).

The value of the constant $C$ increases towards the upper right.

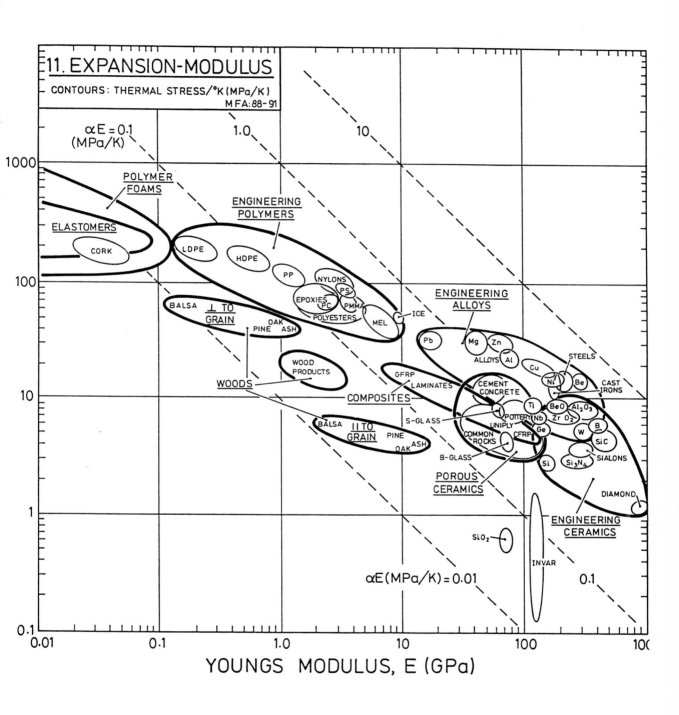

**Materials Selection Chart 12**

**Normalised Strength $\sigma_t/E$, against Linear Expansion Coefficient, $\alpha$**

The chart guides in selecting materials to resist damage in a sudden change of temperature $\Delta T$. The contours show values of the thermal shock parameter $B\Delta T = \dfrac{\sigma_t}{\alpha E}$ in °C. Here $\sigma_t$ is the tensile failure strength (the yield strength of ductile materials, the fracture strength of those which are brittle), $E$ is Young's modulus and $B$ is a factor which allows for constraint and for heat transfer considerations:

$$
\begin{aligned}
B &= & 1/A \ \text{(axial constraint)} \\
&= & (1-\nu)/A \ \text{(biaxial constraint)} \\
&= & (1-2\nu)/A \ \text{(triaxial constraint)}
\end{aligned}
$$

and $\quad A = \dfrac{th/\lambda}{1 + th/\lambda}$

and $\nu$ is Poisson's ratio, $t$ a typical sample dimension, $h$ is the heat transfer coefficient at the sample surface and $\lambda$ is its thermal conductivity. The contours define the guideline

(a)  $B\Delta T = C$ (thermal shock resistance).

The value of the constant $C$ increases towards the top left. The values of $\Delta T$ approaching 500°C for polymers, woods and elastomers do not mean they can be used at 500°C, but that there is a wide safety margin when they are suddenly heated to — say — 100°C.

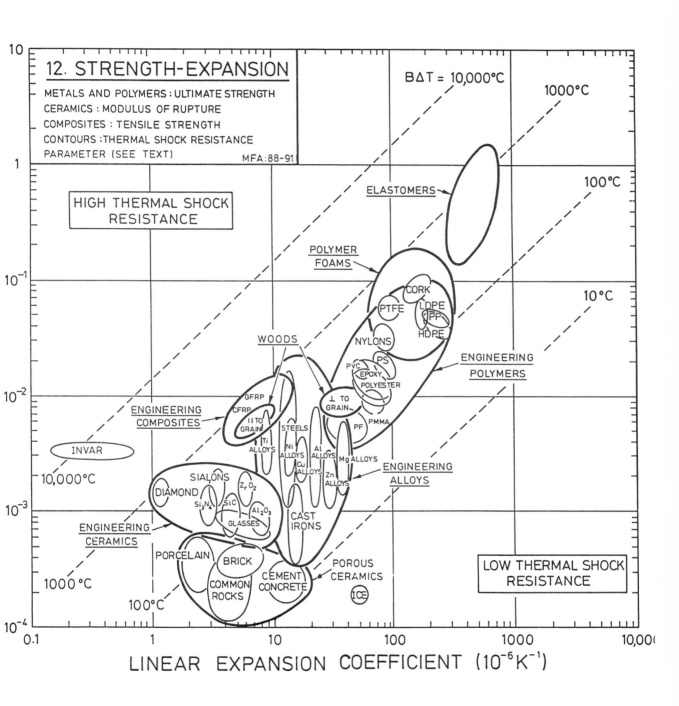

## 12. STRENGTH-EXPANSION

METALS AND POLYMERS : ULTIMATE STRENGTH
CERAMICS : MODULUS OF RUPTURE
COMPOSITES : TENSILE STRENGTH
CONTOURS : THERMAL SHOCK RESISTANCE
PARAMETER (SEE TEXT)
MFA:88-91

$B \Delta T = 10,000°C$

HIGH THERMAL SHOCK RESISTANCE

ELASTOMERS

POLYMER FOAMS

CORK
PTFE   LDPE  PP  HDPE
NYLONS
PS
PVC  EPOXY  POLYESTER
WOODS
GFRP
CFRP
ENGINEERING COMPOSITES
II TO GRAIN
⊥ TO GRAIN
PF   PMMA
ENGINEERING POLYMERS
STEELS
Ti ALLOYS
Ni ALLOYS
Cu ALLOYS
Al ALLOYS
Mg ALLOYS
Zn ALLOYS
ENGINEERING ALLOYS
INVAR
SIALONS
DIAMOND
$Zr O_2$
$Si_3N_4$  SiC  $Al_2O_3$
GLASSES
CAST IRONS
ENGINEERING CERAMICS
PORCELAIN
BRICK
COMMON ROCKS
CEMENT CONCRETE
POROUS CERAMICS
ICE
LOW THERMAL SHOCK RESISTANCE

1000°C
100°C
10°C
10,000°C
1000°C
100°C

LINEAR EXPANSION COEFFICIENT $(10^{-6} K^{-1})$

## Materials Selection Chart 13

### Strength at Temperature, σ(T), against Temperature, T

Materials tend to show a strength which is almost independent of temperature up to a given temperature (the "onset of creep" temperature); above this temperature the strength falls, often steeply. The lozenges show this behaviour (see inset at the bottom right). The "strength" here is a short-term yield strength, corresponding to 1 hour of loading. For long loading times (10,000 hours, for instance), the strengths are lower.

Ths chart gives an overview of high-temperature strength, giving guidance in making an initial choice. Design against creep and creep fracture requires further information and techniques.

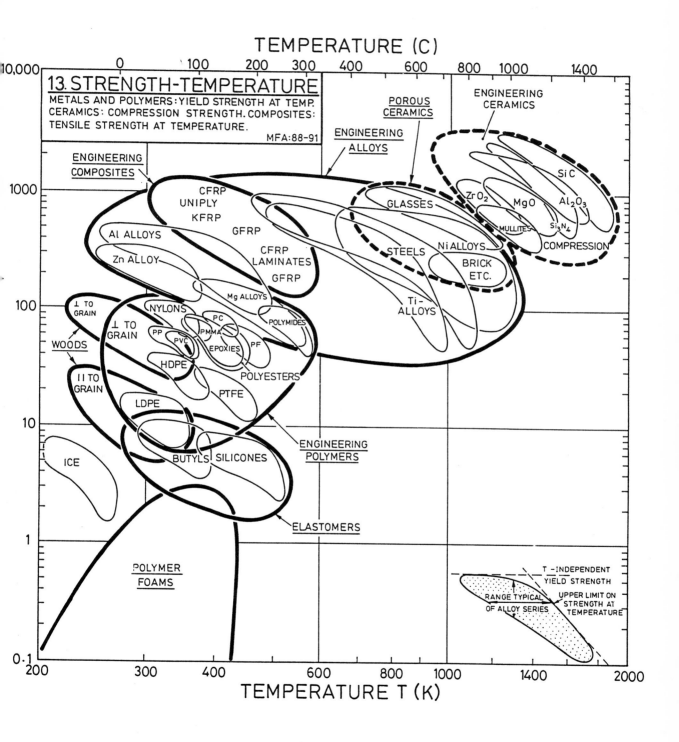

## TEMPERATURE (C)

# 13. STRENGTH-TEMPERATURE
METALS AND POLYMERS: YIELD STRENGTH AT TEMP.
CERAMICS: COMPRESSION STRENGTH. COMPOSITES:
TENSILE STRENGTH AT TEMPERATURE.
MFA:88–91

ENGINEERING
CERAMICS

POROUS
CERAMICS

ENGINEERING
ALLOYS

ENGINEERING
COMPOSITES

CFRP
UNIPLY
KFRP

GFRP

SiC

GLASSES

Zr O₂    MgO    Al₂O₃

Al ALLOYS

CFRP
LAMINATES

MULLITES    Si₃N₄

Zn ALLOY

GFRP

STEELS    Ni ALLOYS    COMPRESSION

Mg ALLOYS

BRICK
ETC.

⊥ TO
GRAIN

NYLONS

PC

POLYMIDES

Ti-
ALLOYS

⊥ TO
GRAIN

PP    PMMA

WOODS

PVC

EPOXIES    PF

HDPE

POLYESTERS

∥ TO
GRAIN

PTFE

LDPE

ICE

BUTYLS SILICONES

ENGINEERING
POLYMERS

ELASTOMERS

POLYMER
FOAMS

T – INDEPENDENT
YIELD STRENGTH

RANGE TYPICAL
OF ALLOY SERIES

UPPER LIMIT ON
STRENGTH AT
TEMPERATURE

10,000

1000

100

10

1

0.1

## TEMPERATURE T (K)

200    300    400    600    800    1000    1400    2000

## Materials Selection Chart 14
### Young's Modulus, E, against Relative Cost, $C_R\rho$

The chart guides selection of materials for cheap, stiff, components (material cost only). The relative cost $C_R$ is calculated by taking that for mild steel reinforcing rods as unity; thus:

$$C_R = \frac{\text{cost per unit weight of the material}}{\text{cost per unit weight of mild steel}}$$

The guidelines show the loci of points for which

   (a)   $E/C_R\rho = C$ (minimum cost design of stiff ties, etc.);
   (b)   $E^{1/2}/C_R\rho = C$ (minimum cost design of stiff beams, shafts and columns);
   (c)   $E^{1/3}/C_R\rho = C$ (minimum cost design of stiff plates).

The value of the constant $C$ increases as the lines are displaced upwards and to the left. Material offering the greatest stiffness per unit cost lie towards the upper left corner. Other moduli are obtained approximately from $E$ by

   (a)   $\nu \approx \frac{1}{3}$; $G \approx \frac{3}{8}E$; $K \approx E$ (metals, ceramics, glasses and glassy polymers);

   (b)   $\nu \approx \frac{1}{2}$; $G \approx \frac{1}{3}E$; $K \approx 10\,E$ (elastomers, rubbery polymers);

where $\nu$ is Poisson's ratio, $G$ is the shear modulus and $K$ is the bulk modulus.

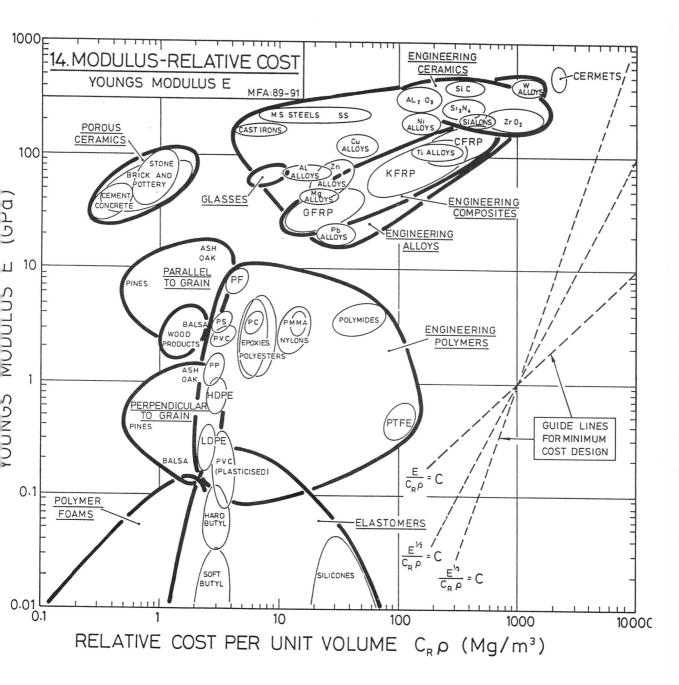

14. MODULUS-RELATIVE COST
YOUNGS MODULUS E
MFA:89-91

YOUNGS MODULUS E (GPa)

RELATIVE COST PER UNIT VOLUME $C_R \rho$ (Mg/m³)

ENGINEERING CERAMICS

CERMETS

W ALLOYS

Si C

AL₂O₃

Si₃N₄

Ni ALLOYS

SIALONS

Zr O₂

MS STEELS

SS

CAST IRONS

CFRP

Cu ALLOYS

Ti ALLOYS

POROUS CERAMICS

STONE

BRICK AND POTTERY

AL ALLOYS

Zn ALLOYS

KFRP

CEMENT CONCRETE

GLASSES

Mg ALLOYS

GFRP

ENGINEERING COMPOSITES

Pb ALLOYS

ENGINEERING ALLOYS

ASH OAK

PARALLEL TO GRAIN

PINES

PF

BALSA WOOD PRODUCTS

PS

PVC

PC

EPOXIES

POLYESTERS

PMMA

NYLONS

POLYMIDES

ENGINEERING POLYMERS

ASH OAK

PP

HDPE

PERPENDICULAR TO GRAIN

PINES

LDPE

BALSA

PVC (PLASTICISED)

PTFE

POLYMER FOAMS

HARD BUTYL

ELASTOMERS

$$\frac{E}{C_R \rho} = C$$

$$\frac{E^{1/2}}{C_R \rho} = C$$

$$\frac{E^{1/3}}{C_R \rho} = C$$

GUIDE LINES FOR MINIMUM COST DESIGN

SOFT BUTYL

SILICONES

## Materials Selection Chart 15

### Strength, $\sigma_f$, against Relative Cost, $C_R\rho$

The "strength" for *metals* is the 0.2% offset yield strength. For *polymers*, it is the 1% offset yield strength. For *ceramics* and *glasses*, it is the compressive crushing strength; remember that this is roughly fifteen times larger than the tensile (fracture) strength. For *composites*, it is the tensile strength. For *elastomers*, it is the tear strength. The relative cost $C_R$ is calculated by taking that of mild steel reinforcing rods as unity; thus

$$C_R = \frac{\text{cost per unit weight of the material}}{\text{cost per unit weight of mild steel}}$$

The chart guides in selecting materials for cheap, strong components (production, finishing assembly and other costs must be considered separately). The lines show the loci of points for which

   (a)   $\sigma_f/C_R\rho = C$ (minimum cost design of strong ties, rotating disks, etc.);
   (b)   $\sigma_f^{2/3}/C_R\rho = C$ (minimum cost design of strong beams and shafts);
   (c)   $\sigma_f^{1/2}/C_R\rho = C$ (minimum cost design of strong plates).

The value of the constants $C$ increase as the lines are displaced upwards and to the left. Materials offering the greatest strength per unit cost lie towards the upper left corner.

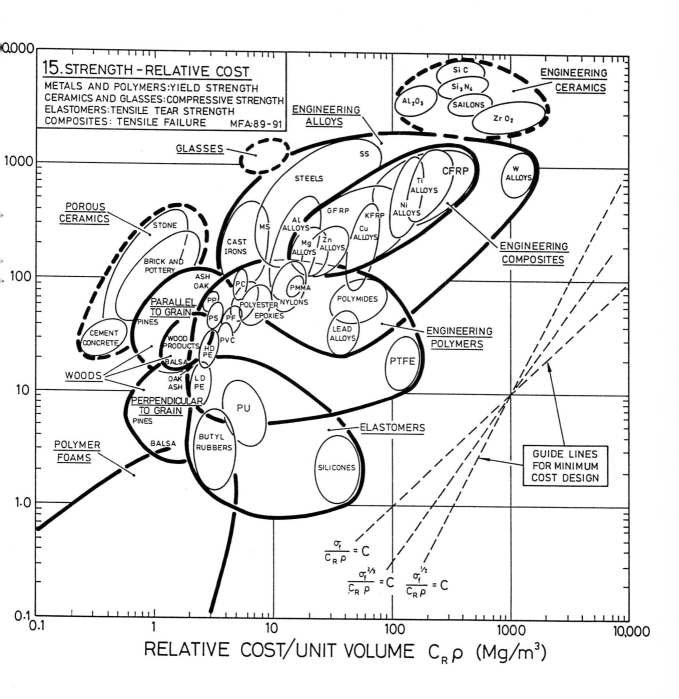

**15. STRENGTH – RELATIVE COST**
METALS AND POLYMERS: YIELD STRENGTH
CERAMICS AND GLASSES: COMPRESSIVE STRENGTH
ELASTOMERS: TENSILE TEAR STRENGTH
COMPOSITES: TENSILE FAILURE          MFA: 89-91

ENGINEERING ALLOYS

GLASSES

ENGINEERING CERAMICS

Si C
Si₃N₄
Al₂O₃
SAILONS
Zr O₂

SS
STEELS
Ti ALLOYS
CFRP
W ALLOYS

POROUS CERAMICS
STONE
MS
AL ALLOYS
GFRP
KFRP
Cu ALLOYS
Ni ALLOYS

BRICK AND POTTERY
CAST IRONS
Mg ALLOYS
Zn ALLOYS

ENGINEERING COMPOSITES

ASH OAK
PC
PMMA
NYLONS
POLYMIDES

PARALLEL TO GRAIN
PINES
PP
PS
PF
POLYESTER
EPOXIES

CEMENT CONCRETE
WOOD PRODUCTS
PVC
HD PE
LEAD ALLOYS

ENGINEERING POLYMERS

BALSA
OAK ASH
LD PE

WOODS

PERPENDICULAR TO GRAIN
PINES
PU

PTFE

POLYMER FOAMS
BALSA
BUTYL RUBBERS

ELASTOMERS

SILICONES

GUIDE LINES FOR MINIMUM COST DESIGN

$$\frac{\sigma_f}{C_R \rho} = C$$

$$\frac{\sigma_f^{2/3}}{C_R \rho} = C \qquad \frac{\sigma_f^{1/2}}{C_R \rho} = C$$

**RELATIVE COST/UNIT VOLUME** $C_R \rho$ **(Mg/m³)**

## Materials Selection Chart 16
## Dry Wear Rate against Maximum Bearing Pressure, $P_{max}$

Wear rate is defined as:

$$W = \frac{\text{volume removed}}{\text{distance slid}}$$

Archard's law, broadly describing wear rates at sliding velocities below 1 m/s, states that

$$W = K_A A P$$

where $A$ is the area, $P$ the bearing pressure (force per unit area) at the sliding surfaces and $K_A$ is Archard's constant. At low bearing pressures $K_A$ is a true constant, but as the maximum bearing pressure is approached it rises steeply (see inset, bottom left). The chart shows Archard's constant,

$$K_A = \frac{W}{AP}$$

plotted against the maximum allowable bearing pressure, $P_{max}$. Bearings cannot be used at or above $P_{max}$ because they seize. The contours show values of the wear rate per unit area

$$M = \frac{W}{A}$$

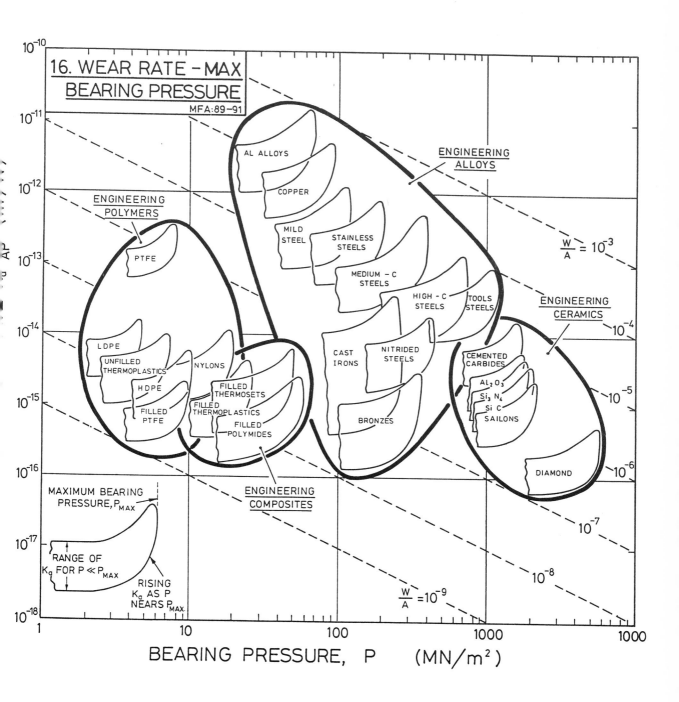

## 16. WEAR RATE — MAX BEARING PRESSURE

MFA:89–91

BEARING PRESSURE, P  (MN/m²)

### Materials Selection Chart 17

### Young's Modulus, E against Energy Content, qρ

The chart guides selection of materials for stiff, energy-economic components. The energy content per m³, $q\rho$, is the energy content per kg, $q$, multiplied by the density, $\rho$. The guidelines show the loci of points for which

(a) $E/q\rho = C$ (minimum energy design of stiff ties; minimum deflection in centrifugal loading, etc.);

(b) $E^{1/2}/q\rho = C$ (minimum energy design of stiff beams, shafts and columns);

(c) $E^{1/3}/q\rho = C$ (minimum energy design of stiff plates).

The value of the constant $C$ increases as the lines are displaced upwards and to the left. Materials offering the greatest stiffness per energy content lie towards the upper left corner. Other moduli are obtained approximately from $E$ using

(a) $\nu \approx \dfrac{1}{3}$; $G \approx \dfrac{3}{8}E$; $K \approx E$ (metals, ceramics, glass and glassy polymers);

(b) $\nu \approx \dfrac{1}{2}$; $G \approx \dfrac{1}{3}E$; $K \approx 10\,E$ (elastomers, rubbery polymers);

where $\nu$ is Poisson's ratio, $G$ the shear modulus and $K$ the bulk modulus.

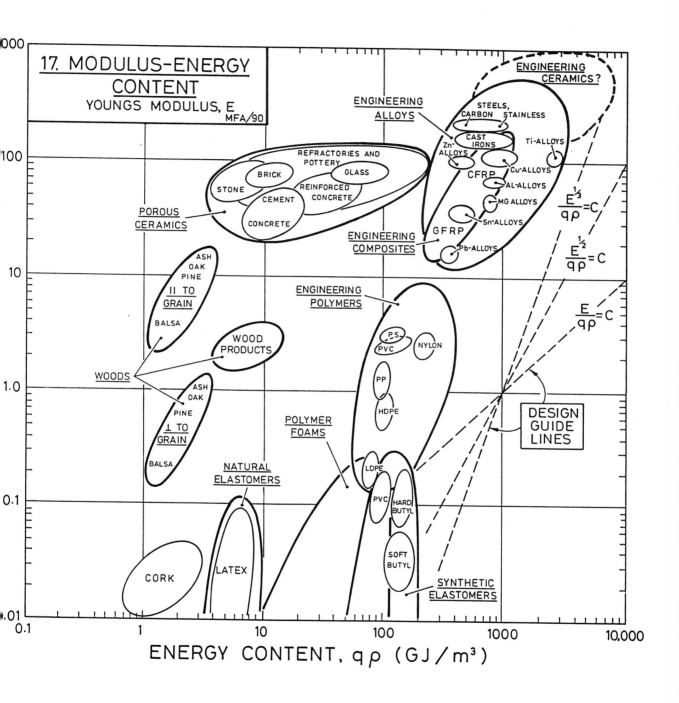

**17. MODULUS–ENERGY CONTENT**
YOUNGS MODULUS, E
MFA/90

ENGINEERING CERAMICS?

ENGINEERING ALLOYS

STEELS, CARBON STAINLESS

CAST IRONS

Zn-ALLOYS

Ti-ALLOYS

CFRP

Cu-ALLOYS

AL-ALLOYS

MG ALLOYS

Sn-ALLOYS

GFRP

Pb-ALLOYS

REFRACTORIES AND POTTERY

BRICK

GLASS

STONE

CEMENT

REINFORCED CONCRETE

CONCRETE

POROUS CERAMICS

ENGINEERING COMPOSITES

ASH OAK PINE

II TO GRAIN

BALSA

ENGINEERING POLYMERS

WOOD PRODUCTS

WOODS

ASH OAK

PINE

⊥ TO GRAIN

BALSA

P.S.

PVC

NYLON

PP

HDPE

POLYMER FOAMS

NATURAL ELASTOMERS

LDPE

PVC HARD BUTYL

SOFT BUTYL

CORK

LATEX

SYNTHETIC ELASTOMERS

$$\frac{E^{\frac{1}{3}}}{q\rho}=C$$

$$\frac{E^{\frac{1}{2}}}{q\rho}=C$$

$$\frac{E}{q\rho}=C$$

DESIGN GUIDE LINES

**ENERGY CONTENT, $q\rho$ (GJ/m³)**

## Materials Selection Chart 18

### Strength $\sigma_f$, against Energy Content, $q\rho$

The "strength" for *metals* is the 0.2% offset yield strength. For *polymers*, it is the stress at which the stress – strain curve becomes markedly nonlinear — typically, a strain of about 1%. For *ceramics and glasses*, it is the compressive crushing strength; remember that this is roughly fifteen times larger than the tensile (fracture) strength. For *composites*, it is the tensile strength. For *elastomers*, it is the tear strength. The energy content per m$^3$, $q\rho$, is the energy content per kg, $q$, multiplied by the density, $\rho$. The chart guides selection of materials for strong, energy-economic components. The guidelines show the loci of points for which

(a)  $\sigma_f/q\rho$ $\;=\;C$ (minimum energy design of strong ties; maximum rotational velocity of disks);

(b)  $\sigma_f^{2/3}/q\rho = C$ (minimum energy design of strong beams and shafts);

(c)  $\sigma_f^{1/2}/q\rho = C$ (minimum energy design of strong plates).

The value of the constant $C$ increases as the lines are displaced upwards and to the left. Materials offering the greatest strength per unit energy content lie towards the upper left corner.

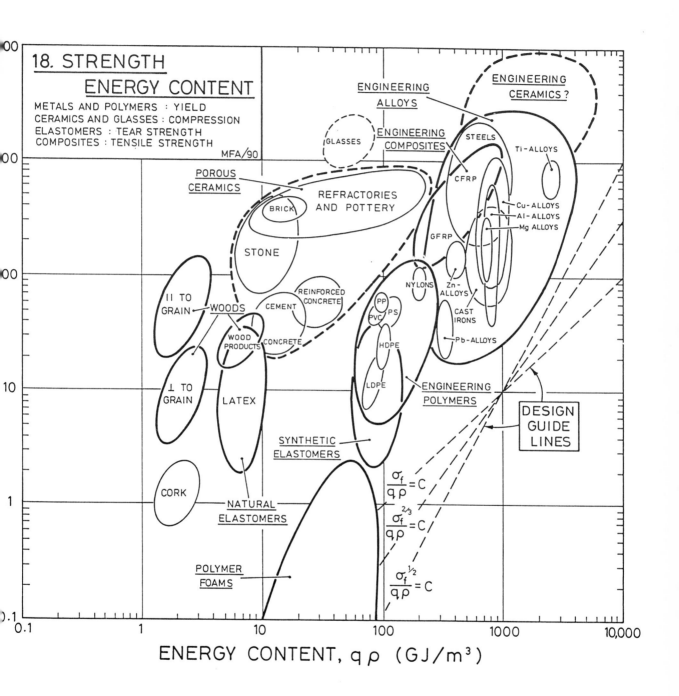

**18. STRENGTH ENERGY CONTENT**

METALS AND POLYMERS : YIELD
CERAMICS AND GLASSES : COMPRESSION
ELASTOMERS : TEAR STRENGTH
COMPOSITES : TENSILE STRENGTH

MFA/90

ENGINEERING ALLOYS

ENGINEERING CERAMICS ?

GLASSES

ENGINEERING COMPOSITES

STEELS

Ti-ALLOYS

POROUS CERAMICS

REFRACTORIES AND POTTERY

BRICK

CFRP

Cu-ALLOYS
Al-ALLOYS
Mg ALLOYS

STONE

GFRP

REINFORCED CONCRETE

II TO GRAIN — WOODS

CEMENT

NYLONS

Zn-ALLOYS

PP
PVC PS

CAST IRONS

WOOD PRODUCTS

CONCRETE

HDPE

Pb-ALLOYS

⊥ TO GRAIN

LATEX

LDPE

ENGINEERING POLYMERS

DESIGN GUIDE LINES

SYNTHETIC ELASTOMERS

$$\frac{\sigma_f}{q\rho} = C$$

CORK

NATURAL ELASTOMERS

$$\frac{\sigma_f^{2/3}}{q\rho} = C$$

POLYMER FOAMS

$$\frac{\sigma_f^{1/2}}{q\rho} = C$$

**ENERGY CONTENT, q ρ (GJ/m³)**

0.1     1     10     100     1000     10,000

# C5 The Process Selection Charts

### Process Selection Chart P1

### Surface Area, A, against Minimum Section, t

This chart (and the next two which show details) guides selection of process when size or shape are limiting. The volume, $V$, for uniform sections is, within a factor of 2, given by

$$C = At$$

Contours of volume are shown as a family of diagonal lines. Volume is converted approximately to weight using an "average" material density of 5000 kg/m$^3$ — most engineering materials have densities within a factor of 2 of this value (polymers are the exception — all have densities near 1000 kg/m$^3$).

The slenderness ratio, $\lambda$, is measured approximately by $t/A^{1/2}$. Contours of this slenderness run diagonally from top right to bottom left. The bottom right corner of the $A - t$ space is inaccessible: no real shapes lie here.

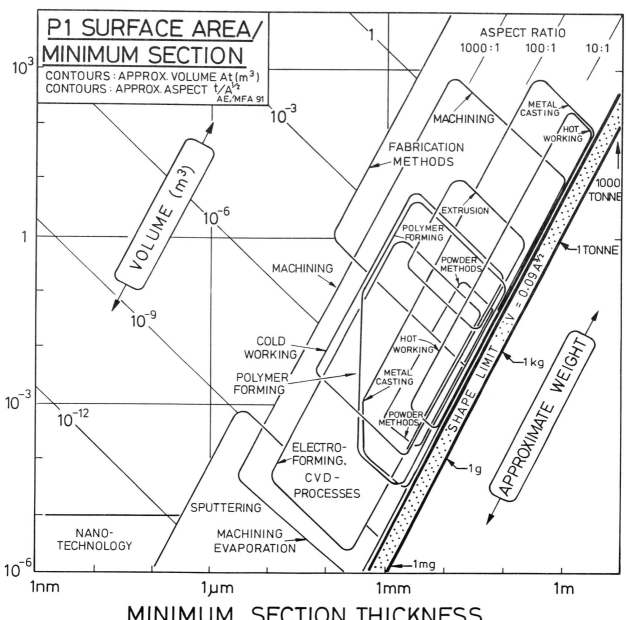

## MINIMUM SECTION THICKNESS

### Process Selection Chart P1(a)

### Surface Area, A, against Minimum Section, t

This chart shows details of part of Chart P1 for casting and vapour-forming processes. It guides selection of process when size or shape are limiting. The volume, $V$, for uniform sections is, within a factor of 2, given by

$$V = At$$

Contours of volume are shown as a family of diagonal lines. Volume is converted approximately to weight using an "average" material density of 5000 kg/m$^3$ — most engineering materials have densities within a factor of 2 of this value (polymers are the exception — all have densities near 1000 kg/m$^3$).

The slenderness ratio, $\lambda$, is measured approximately by $t/A^{1/2}$. Contours of this slenderness run diagonally from top right to bottom left. The bottom right corner of the $A - t$ space is inaccessible: no real shapes lie here.

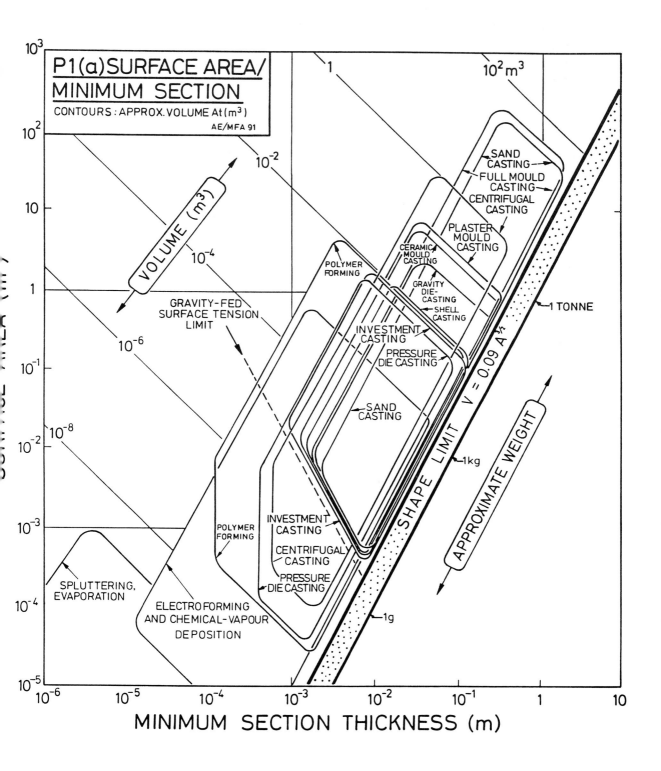

P1(a) SURFACE AREA/ MINIMUM SECTION
CONTOURS : APPROX. VOLUME At (m³)
AE/MFA 91

MINIMUM SECTION THICKNESS (m)

### Process Selection Chart P1(b)
### Surface Area, A, against Minimum Section, t

This chart shows details of part of Chart P1 for deformation and polymer-forming processes. It guides selection of process when size or shape are limiting. The volume, $V$, for uniform sections is, within a factor of 2, given by

$$V = At$$

Contours of volume are shown as a family of diagonal lines. Volume is converted approximately to weight using an "average" material density of 5000 kg/m$^3$ — most engineering materials have densities within a factor of 2 of this value (polymers are the exception — all have densities near 1000 kg/m$^3$).

The slenderness ratio, $\lambda$, is measured approximately by $t/A^{1/2}$. Contours of this slenderness run diagonally from top right to bottom left. The bottom right corner of the $A - t$ space is inaccessible: no real shapes lie here.

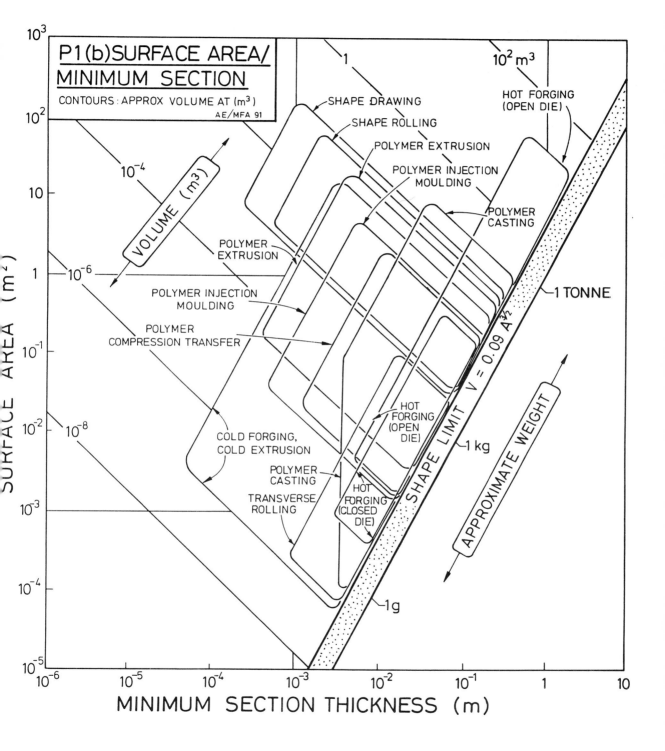

P1(b) SURFACE AREA/ MINIMUM SECTION. CONTOURS: APPROX VOLUME AT (m³). AE/MFA 91. Chart plotting SURFACE AREA (m²) against MINIMUM SECTION THICKNESS (m), showing process regions: SHAPE DRAWING, SHAPE ROLLING, POLYMER EXTRUSION, POLYMER INJECTION MOULDING, POLYMER CASTING, HOT FORGING (OPEN DIE), HOT FORGING (CLOSED DIE), COLD FORGING COLD EXTRUSION, POLYMER COMPRESSION TRANSFER, TRANSVERSE ROLLING. Reference lines: VOLUME (m³), SHAPE LIMIT $V = 0.09\,A^{3/2}$, APPROXIMATE WEIGHT (1g, 1kg, 1 TONNE).

## Process Selection Chart P2
## Complexity, C, against Size, W

This chart guides in selection of process when complexity or size are limiting. Complexity, measured in bits, is defined by the quantity.

$$C = n \log_2 \left( \frac{\bar{\ell}}{\overline{\Delta \ell}} \right)$$

where $n$ is the number of the dimensions, and $\overline{\Delta \ell}/\bar{\ell}$ the (average) fractional precision with which they are specified. The lower limit of complexity is $C \approx 1$: a sphere, with only one dimension, specified to a tolerance of $\pm 50\%$. There is no obvious upper limit: fabrication (that is, assembly and joining) allows complexity to be increased without limit. In practice, reliability sets an upper practical limit.

Size is measured by the weight, $W$. Volume $V$ can be calculated from this via the density.

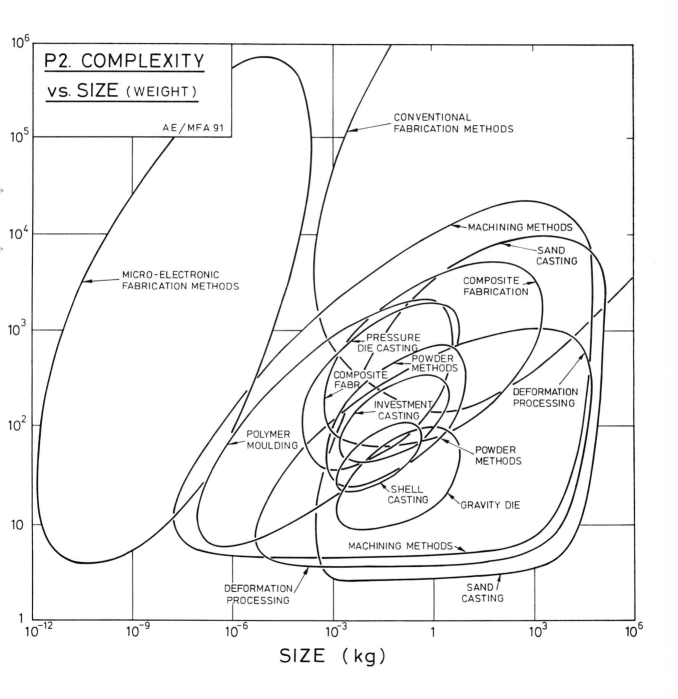

P2. COMPLEXITY vs. SIZE (WEIGHT)

AE/MFA 91

CONVENTIONAL FABRICATION METHODS

MACHINING METHODS

SAND CASTING

COMPOSITE FABRICATION

MICRO-ELECTRONIC FABRICATION METHODS

PRESSURE DIE CASTING

POWDER METHODS

COMPOSITE FABR

DEFORMATION PROCESSING

INVESTMENT CASTING

POLYMER MOULDING

POWDER METHODS

SHELL CASTING

GRAVITY DIE

MACHINING METHODS

DEFORMATION PROCESSING

SAND CASTING

SIZE (kg)

### Process Selection Chart P3

### Size, W, against Melting Temperature, $T_m$

This chart and the next guide in matching process to material. The choice of casting, moulding or vapour-forming process is greatly influenced by melting point: low-melting materials can be cast by many alternative routes: high-melting materials by only a few — for these, powder or vapour phase methods may be necessary. The size is measured here by the weight, $W$. It can be converted to volume via the density, $\rho$, or to the approximate linear dimension, $L$, shown on the right-hand axis via

$$L = (\frac{W}{\rho})^{1/3}$$

The melting point is in Kelvin.

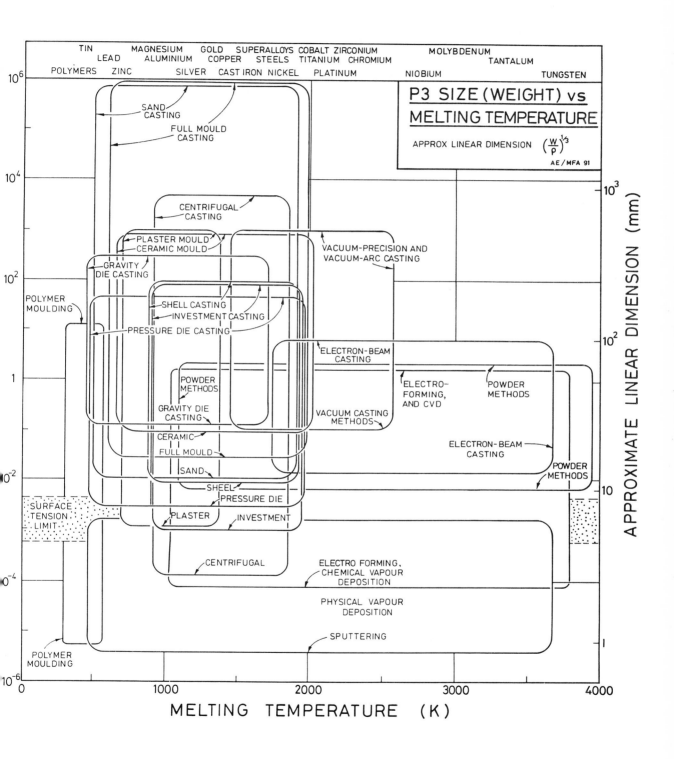

### Process Selection Chart P4

### Hardness, H, against Melting Temperature, $T_m$

This chart and the last guide in matching process to material. Casting and moulding techniques are limited by melting temperature. Deformation processing methods are limited by flow strength (or, equivalently, hardness) and ductility. The flow strength is measured here by the room temperature hardness, $H$. It is plotted against melting temperature, $T_m$, in Kelvin. The two quantities are not independent, but, for real materials, are bracketted by the relation

$$0.03 < \frac{H\Omega}{kT_m} < 20$$

where $\Omega$ is the atomic or molecular volume and $k$ is Boltzmann's Constant. The two limits are shown as heavy black lines. Real materials have hardnesses and melting temperatures which lie between these limits.

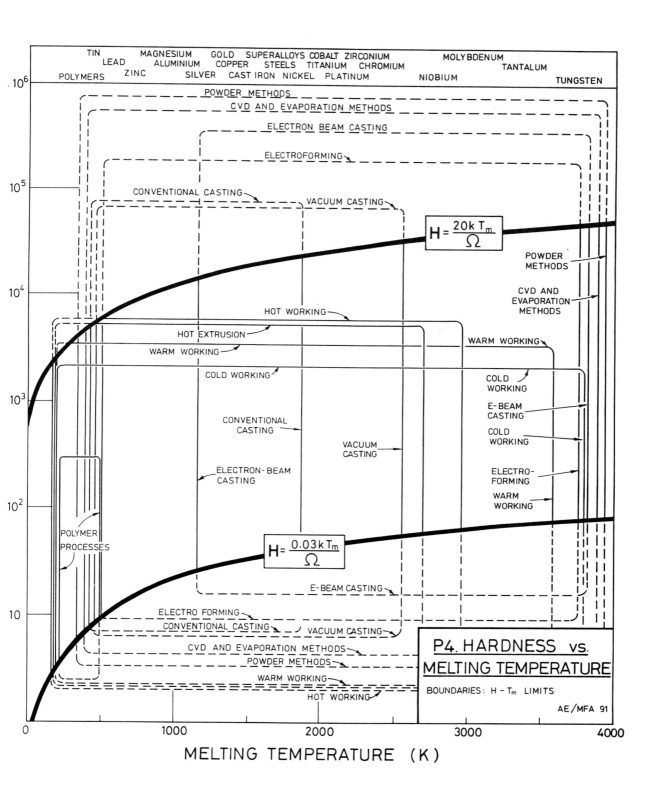

**P4. HARDNESS vs. MELTING TEMPERATURE**

BOUNDARIES: H – $T_m$ LIMITS

AE/MFA 91

MELTING TEMPERATURE (K)

## Process Selection Chart P5

### Tolerance Range, T, against RMS Surface Roughness, R

This chart guides in selecting processes which meet specifications on tolerance and roughness. The tolerance $T$ is the permitted slack in the dimension of a part: it is specified as $100 \pm 0.1$ mm or as $50\,^{+0.01}_{-0.001}$ mm. The surface finish $R$ is measured by the root mean square amplitude of the irregularities on the surface: it is specified as $R = 10$ $\mu$m (the rough surface of a sand casting) or $R = 0.05$ $\mu$m (a lapped surface).

The tolerance must be greater than $2R$ (shaded band): since $R$ is the root mean square roughness, the highest peaks of the irregularities are about $5R$ in height. Real processes give tolerances which range from about $10R$ to $1000R$.

Processing costs increase almost exponentially as the requirements for tolerance and surface finish are made more severe. The contours show the relative cost: an increase in precision corresponding to the separation of two neighbouring contours gives an increase in cost, for a given process, of a factor of about 2.

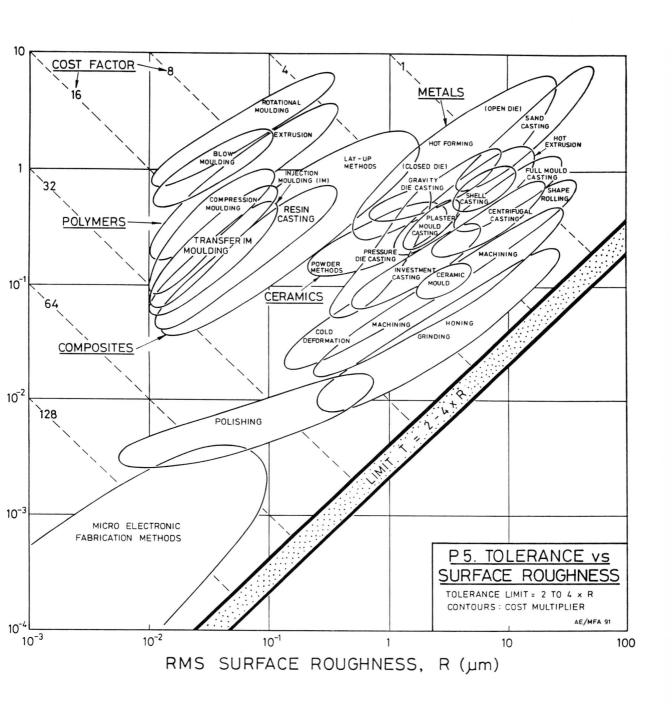

COST FACTOR 8

16

4

1

32

METALS

ROTATIONAL MOULDING

(OPEN DIE)

SAND CASTING

EXTRUSION

HOT EXTRUSION

BLOW MOULDING

HOT FORMING

LAY-UP METHODS

FULL MOULD CASTING

(CLOSED DIE)

INJECTION MOULDING (IM)

GRAVITY DIE CASTING

SHAPE ROLLING

COMPRESSION MOULDING

SHELL CASTING

RESIN CASTING

POLYMERS

CENTRIFUGAL CASTING

PLASTER MOULD CASTING

TRANSFER IM MOULDING

64

PRESSURE DIE CASTING

POWDER METHODS

MACHINING

COMPOSITES

INVESTMENT CASTING

CERAMIC MOULD

CERAMICS

MACHINING

HONING

COLD DEFORMATION

GRINDING

128

POLISHING

LIMIT T = 2 - 4 × R

MICRO ELECTRONIC FABRICATION METHODS

**P 5. TOLERANCE vs SURFACE ROUGHNESS**

TOLERANCE LIMIT = 2 TO 4 × R
CONTOURS : COST MULTIPLIER

AE/MFA 91

RMS SURFACE ROUGHNESS, R (μm)